Bond
No.1 for exam success

SATs Skills

Times Tables Workbook

for Key Stage 2

OXFORD
UNIVERSITY PRESS

Great Clarendon Street, Oxford, OX2 6DP, United Kingdom

Oxford University Press is a department of the University of Oxford. It furthers the University's objective of excellence in research, scholarship, and education by publishing worldwide. Oxford is a registered trade mark of Oxford University Press in the UK and in certain other countries.

First published in 2016

British Library Cataloguing in Publication Data
Data available

ISBN: 978-0-19-274568-2

10 9 8 7 6 5 4 3 2

Paper used in the production of this book is a natural, recyclable product made from wood grown in sustainable forests. The manufacturing process conforms to the environmental regulations of the country of origin.

Printed in China

Acknowledgements

Cover illustration: Lo Cole
llustrations by Lo Cole

Although we have made every effort to trace and contact all copyright holders before publication this has not be possible in all cases. If notified the publisher will rectify any error or omissions at the earliest opportunity.

Test your skills: the 2, 3, 5 and 10 times tables

Pattern alert!

Knowing the patterns can help you remember your **times tables**. Look out for them.

The answers to the 2 **times table** are even numbers and end in 2, 4, 6, 8 or 0.

The answers to the 5 **times table** end in 5 or 0.

The answers to the 10 **times table** always end in 0.

Ⓐ How well do you remember these times tables?
How quickly can you fill in the missing numbers? [16]

2 × table

3 × 2 = ______

1 × 2 = ______

______ × 2 = 14

4 × ______ = 8

______ × 2 = 20

5 × 2 = ______

1 × 2 = ______

______ × 2 = 12

3 × table

7 × 3 = ______

______ × 3 = 30

3 × ______ = 12

______ × 3 = 27

12 × 3 = ______

3 × ______ = 33

3 × 3 = ______

______ × 3 = 3

Total marks

/16

(B) How well do you remember these times tables?
How quickly can you fill in the missing numbers? [16]

5× table

7 × 5 = ______

______ × 5 = 10

4 × ______ = 20

______ × 5 = 45

12 × 5 = ______

______ × 5 = 40

3 × 5 = ______

11 × ______ = 55

10× table

5 × 10 = ______

3 × 10 = ______

8 × ______ = 80

______ × 10 = 60

11 × 10 = ______

4 × ______ = 40

______ × 10 = 20

12 × 10 = ______

Helpful Hint

Multiplying a number by 2 is the same as **doubling** it.

Multiplying by 5 is the same as multiplying by 10 and then **halving**.

Multiplications can be written in any order and the answer will still be the same.

Example: 5 × 2 = 10
2 × 5 = 10

Total marks

/16

Helpful Hint

Multiples are the result of multiplying one number with another.

Example: Multiples of 5 are **5** (1 × 5), **10** (2 × 5), **15** (3 × 5), **20** (4 × 5) and so on.

To find the **product** of two or more numbers, you **multiply** them together. If you **multiply** a number by 1, it is itself. If you **multiply** a number by 0, the answer is 0.

Your **times tables** can also help you **divide**.

4 groups of 5 is 20 — 20 shared between 5 is 4

4 multiplied by 5 = 20 — 20 divided by 5 = 4

C Circle all the multiples of 2 in the first row of numbers. Then circle all the multiples of 5 in the second row, and all the multiples of 10 in the third row. [15]

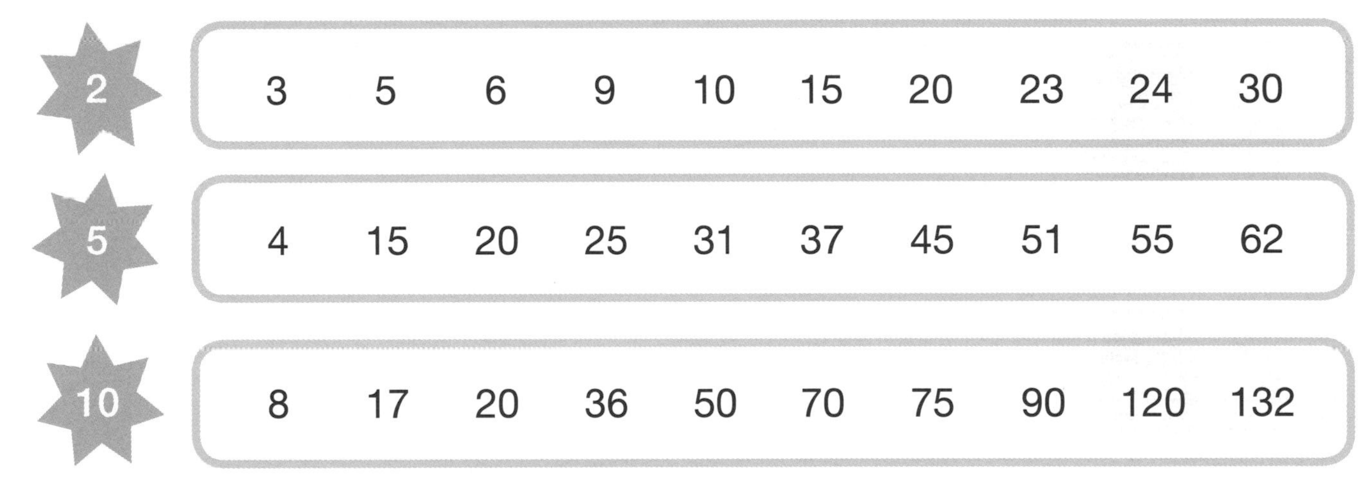

Is there a number that you have circled three times?
This is a multiple of 2, 5 and 10! [1]

D Answer these questions as quickly as possible. [8]

1 What is 5 times 10?

2 Multiply 8 by 10.

3 Find the product of 3 and 2.

4 Share 45 equally between 5.

5 What are 2 lots of 7?

6 Divide 55 by 11.

7 Multiply 10 by itself.

8 How many 6s are there in 30?

Total marks /24

Ⓔ Match the multiplications or divisions to their answers.
Some answers will have more than one match. One has been done for you. [14]

2 × 5

24 ÷ 2

6 × 3

3 × 2

60 ÷ 6

110 ÷ 10

16 ÷ 8

9 × 10

40 ÷ 10

10 × 10

2 × 10

6 × 5

5 × 5

40 ÷ 5

45 ÷ 9

12 18 11 2 10 6 100 90 20 30 4 25 5 8

Total marks

/14

The 4 times table

Pattern alert!

Numbers in the **4 times table** are even and end in either 2, 4, 6, 8 or 0, which is the same pattern as for the **2 times table**.

The **multiples** in the **4 times table** are **double** the **multiples** in the **2 times table.**

To **multiply** a number by 4, **double** it and then **double** it again.

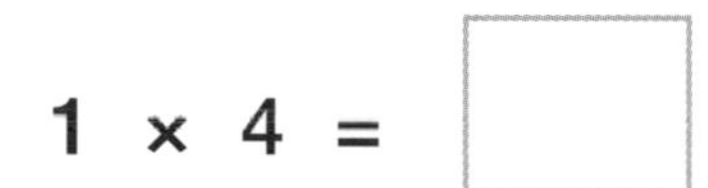

Ⓐ Read and learn the 4 times table at the edge of the page. Now fold it under and fill in the missing numbers. Check your answers. [12]

1 × 4 = ☐	7 × 4 = ☐
2 × 4 = ☐	8 × 4 = ☐
3 × 4 = ☐	9 × 4 = ☐
4 × 4 = ☐	10 × 4 = ☐
5 × 4 = ☐	11 × 4 = ☐
6 × 4 = ☐	12 × 4 = ☐

Ⓑ Can you count in multiples of 4? Fill in the missing numbers. [5]

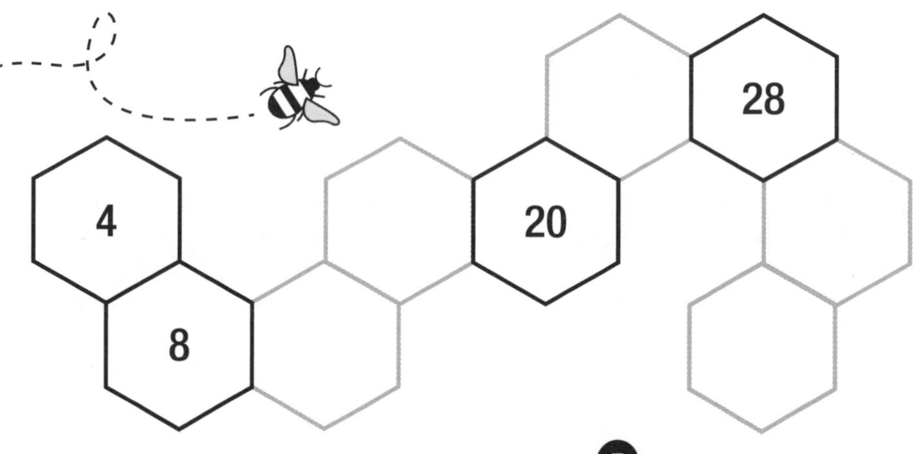

4
times table

1 × 4 = 4

2 × 4 = 8

3 × 4 = 12

4 × 4 = 16

5 × 4 = 20

6 × 4 = 24

7 × 4 = 28

8 × 4 = 32

9 × 4 = 36

10 × 4 = 40

11 × 4 = 44

12 × 4 = 48

Total marks

/ 17

4
times table

1 × 4 = 4

2 × 4 = 8

3 × 4 = 12

4 × 4 = 16

5 × 4 = 20

6 × 4 = 24

7 × 4 = 28

8 × 4 = 32

9 × 4 = 36

10 × 4 = 40

11 × 4 = 44

12 × 4 = 48

Total marks

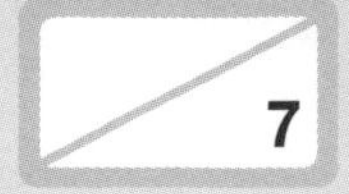

Helpful Hint

The 4× **table** helps you find groups of **4**.

7 × 4 = 28 is the same as 7 'lots of' 4 = 28

7 'groups of' 4 = 28

7 'multiplied by' 4 = 28

(C) Fold back the edge of this page and balance the scales with the correct answers. One has been done for you. [7]

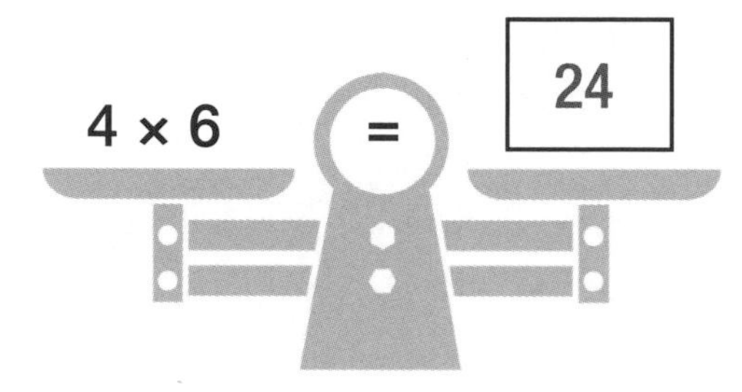

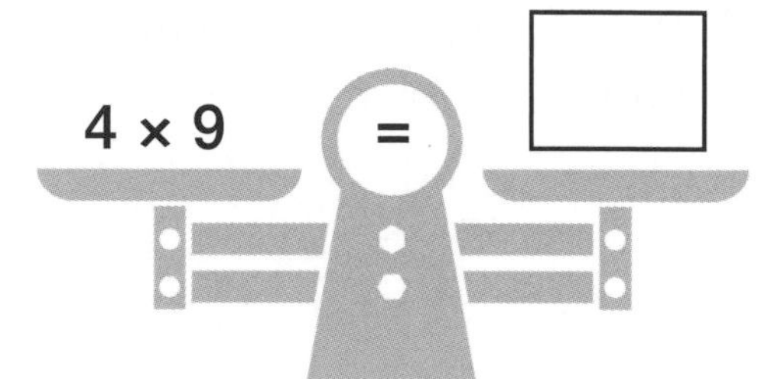

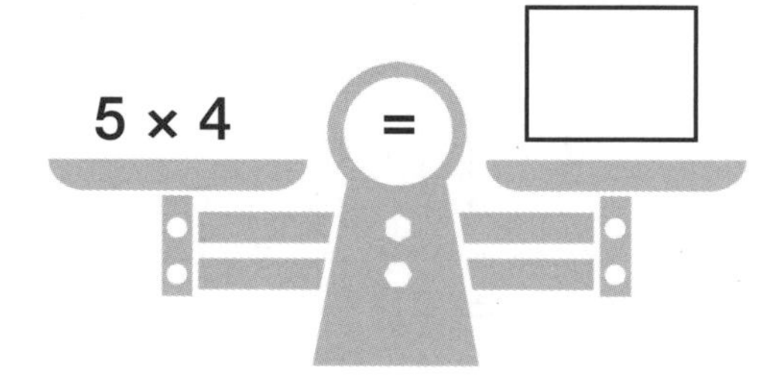

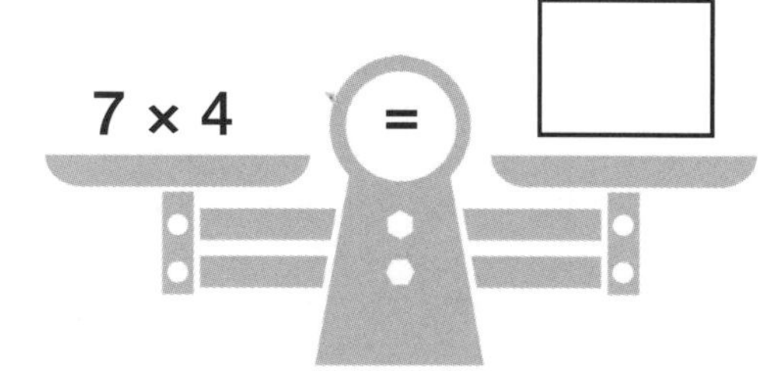

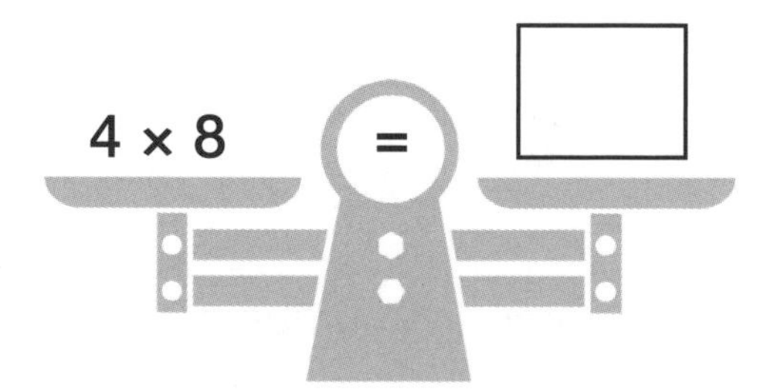

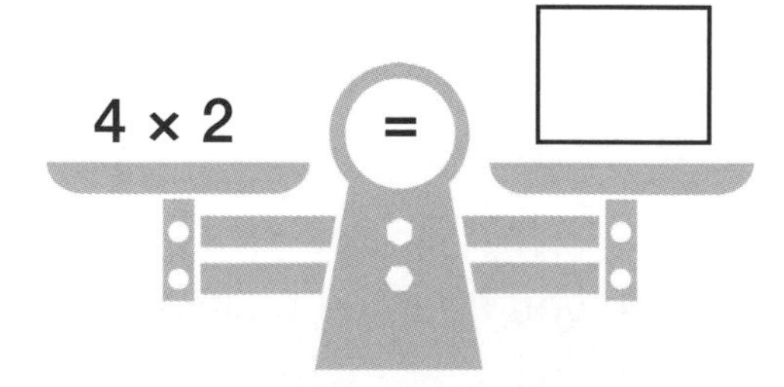

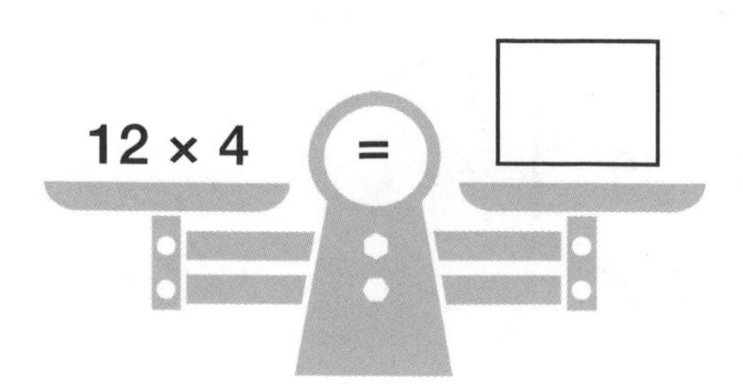

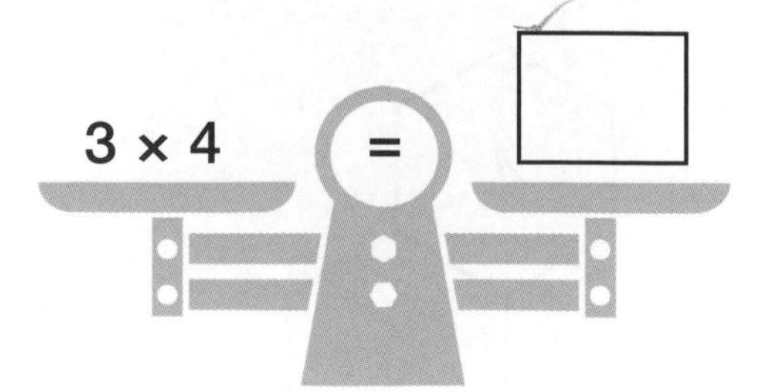

(D) Use the 4 times table to answer these word problems. [3]

1 Willow class is split into four teams of seven. How many children are in the class?

2 Mum bought five packs of cat food. Each pack contains four tins.
How many tins of cat food are there altogether?

3 A pizza is cut into four slices. How many slices would there be if eleven pizzas were cut in the same way?

Helpful Hint

Did you know that **times tables** can help when dividing numbers?

7 × 4 = 28	28 ÷ 4 = 7
7 lots of 4 = 28	28 shared by 4 = 7
7 multiplied by 4 = 28	28 divided by 4 = 7

(E) Put the numbers in the number machines. Write the numbers that come out. One has been done for you. [7]

Total marks

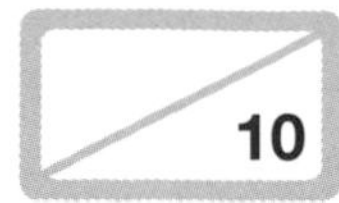

(F) Fill in the missing numbers. [18]

5 × 4 = ☐	24 ÷ ☐ = 4
☐ ÷ 4 = 12	4 × 0 = ☐
☐ × 7 = 28	20 ÷ 4 = ☐
36 ÷ ☐ = 9	☐ × 4 = 16
☐ × 4 = 32	10 × 4 = ☐
8 ÷ 4 = ☐	☐ ÷ 4 = 3
4 ÷ 4 = ☐	4 ÷ ☐ = 4
☐ × 4 = 36	4 × ☐ = 12
11 × 4 = ☐	4 × 6 = ☐

Total marks

/18

The 8 times table

Pattern alert!

The **8 times table** is another table with answers that are even numbers.

You already know these six **facts** from the **8 times table** from other tables:

1 × 8 = 8

2 × 8 = 16 (because 2 × 8 is the same as 8 × 2)

3 × 8 = 24 (because 3 × 8 is the same as 8 × 3)

4 × 8 = 32 (because 4 × 8 is the same as 8 × 4)

5 × 8 = 40 (half of 10 × 8)

10 × 8 = 80 (8 with a zero added to the end).

The **4 times table** can also help you with the **8 times table** because the **8 times table** is double the **4 times table**!

Example: 3 × 4 = 12
3 × 8 = 24

8
times table

1 × 8 = 8

2 × 8 = 16

3 × 8 = 24

4 × 8 = 32

5 × 8 = 40

6 × 8 = 48

7 × 8 = 56

8 × 8 = 64

9 × 8 = 72

10 × 8 = 80

11 × 8 = 88

12 × 8 = 96

A) Read and learn the 8 times table at the edge of the page. Now cover it and fill in the missing numbers. Check your answers. [12]

1 × 8 = ☐

2 × 8 = ☐

3 × 8 = ☐

4 × 8 = ☐

5 × 8 = ☐

6 × 8 = ☐

7 × 8 = ☐

8 × 8 = ☐

9 × 8 = ☐

10 × 8 = ☐

11 × 8 = ☐

12 × 8 = ☐

Total marks

12

8 times table

1 × 8 = 8
2 × 8 = 16
3 × 8 = 24
4 × 8 = 32
5 × 8 = 40
6 × 8 = 48
7 × 8 = 56
8 × 8 = 64
9 × 8 = 72
10 × 8 = 80
11 × 8 = 88
12 × 8 = 96

Total marks

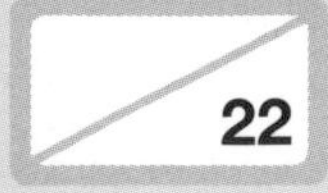

Ⓑ Fold back the edge of this page and fill in the missing numbers. [8]

24 = ☐ × 8	88 = ☐ × 11
80 = ☐ × 10	48 = ☐ × 8
8 = ☐ × 1	32 = ☐ × 4
96 = ☐ × 8	56 = ☐ × 8

Ⓒ Match the numbers to the multiplications or divisions to fill in the missing numbers. [14]

64	6	12	11	24	16	32
40	72	1	96	56	7	8

3 × 8 = ☐	88 ÷ 8 = ☐
8 × 8 = ☐	8 × 4 = ☐
24 ÷ 3 = ☐	2 × 8 = ☐
12 × 8 = ☐	48 ÷ 8 = ☐
5 × 8 = ☐	8 × 7 = ☐
8 ÷ 8 = ☐	96 ÷ 8 = ☐
56 ÷ 8 = ☐	9 × 8 = ☐

Ⓓ Use the 8 times table to answer these word problems. [5]

1 A spider has eight legs. If there are 56 spider legs, how many spiders are there?

____

2 16 apples were collected from an apple tree. The apples were shared equally between two children. How many apples did each child get?

____

3 Five children are given eight different words to write on a display. How many new words are added to the display in total?

____

4 There are eight ducks swimming on the lake. How many wings are there in total?

____

5 Joe has 12 boxes of pens. There are 8 pens in each box. How many pens does he have altogether?

____

Ⓔ Circle all the multiples of 4 in the first row of numbers.
Circle all the multiples of 8 in the second row of numbers. [10]

12 15 16 20 27 36 38 45 48 55

16 19 20 28 32 40 56 75 78 88

Can you find any common multiples of both 4 and 8? [1]

____

Total marks
/16

Ⓕ Put the numbers in the number machines. Write the numbers that come out. [12]

Total marks

/12

The 6 times table

Pattern alert!

All the answers in the **6 times table** are even numbers – they end with 0, 2, 4, 6 or 8.

Also, if you add the **digits** that make the answer they add up to 3, 6 or 9 (except for 11 × 6, where both **digits** are 6 anyway).

The answers to the **6 times table** are **double** that of the **3 times table**.

Ⓐ Read and learn the 6 times table at the edge of the page. Now cover it and fill in the missing numbers. Check your answers. [12]

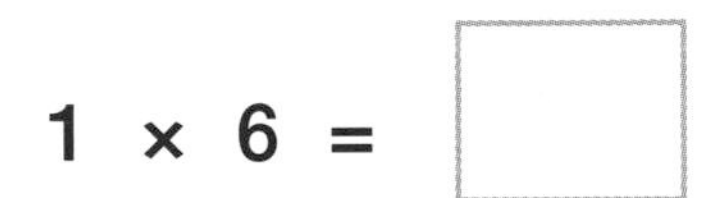

1 × 6 = ☐	7 × 6 = ☐
2 × 6 = ☐	8 × 6 = ☐
3 × 6 = ☐	9 × 6 = ☐
4 × 6 = ☐	10 × 6 = ☐
5 × 6 = ☐	11 × 6 = ☐
6 × 6 = ☐	12 × 6 = ☐

Ⓑ Can you count in multiples of 6? Fill in the missing numbers. [5]

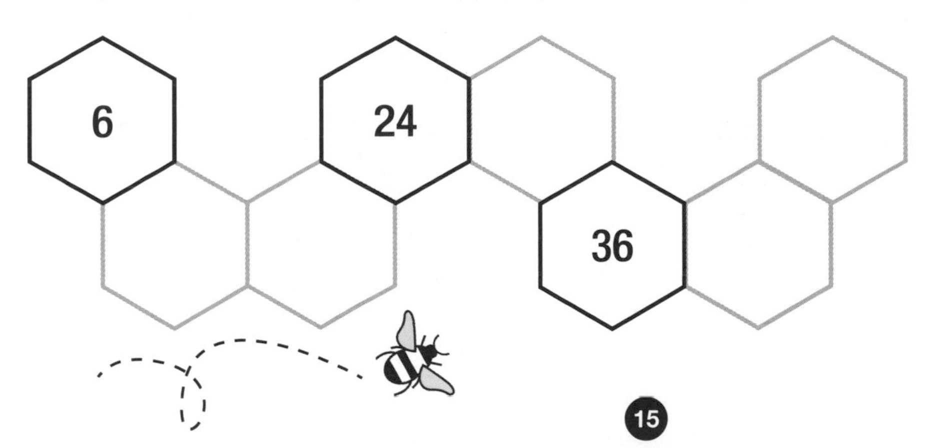

6
times table

1 × 6 = 6

2 × 6 = 12

3 × 6 = 18

4 × 6 = 24

5 × 6 = 30

6 × 6 = 36

7 × 6 = 42

8 × 6 = 48

9 × 6 = 54

10 × 6 = 60

11 × 6 = 66

12 × 6 = 72

Total marks

6
times table

1 × 6 = 6

2 × 6 = 12

3 × 6 = 18

4 × 6 = 24

5 × 6 = 30

6 × 6 = 36

7 × 6 = 42

8 × 6 = 48

9 × 6 = 54

10 × 6 = 60

11 × 6 = 66

12 × 6 = 72

Total marks

Helpful Hint

The **6 times table** helps you find groups of 6.

3 × 6 = 18 is the same as 3 'lots of' 6 = 18

3 'groups of' 6 = 18

3 'multiplied by' 6 = 18

Multiplications can be written in any order and the answer will still be the same.

Example: 3 × 6 = 18 6 × 3 = 18

Remember that **times tables** can help you to **divide**. Ask yourself 'how many of the smaller number are there in the larger number?'

Example: 3 × 6 = 18 18 ÷ 6 = 3

Ⓒ Fold back the edge of this page and complete the honeycomb. [9]

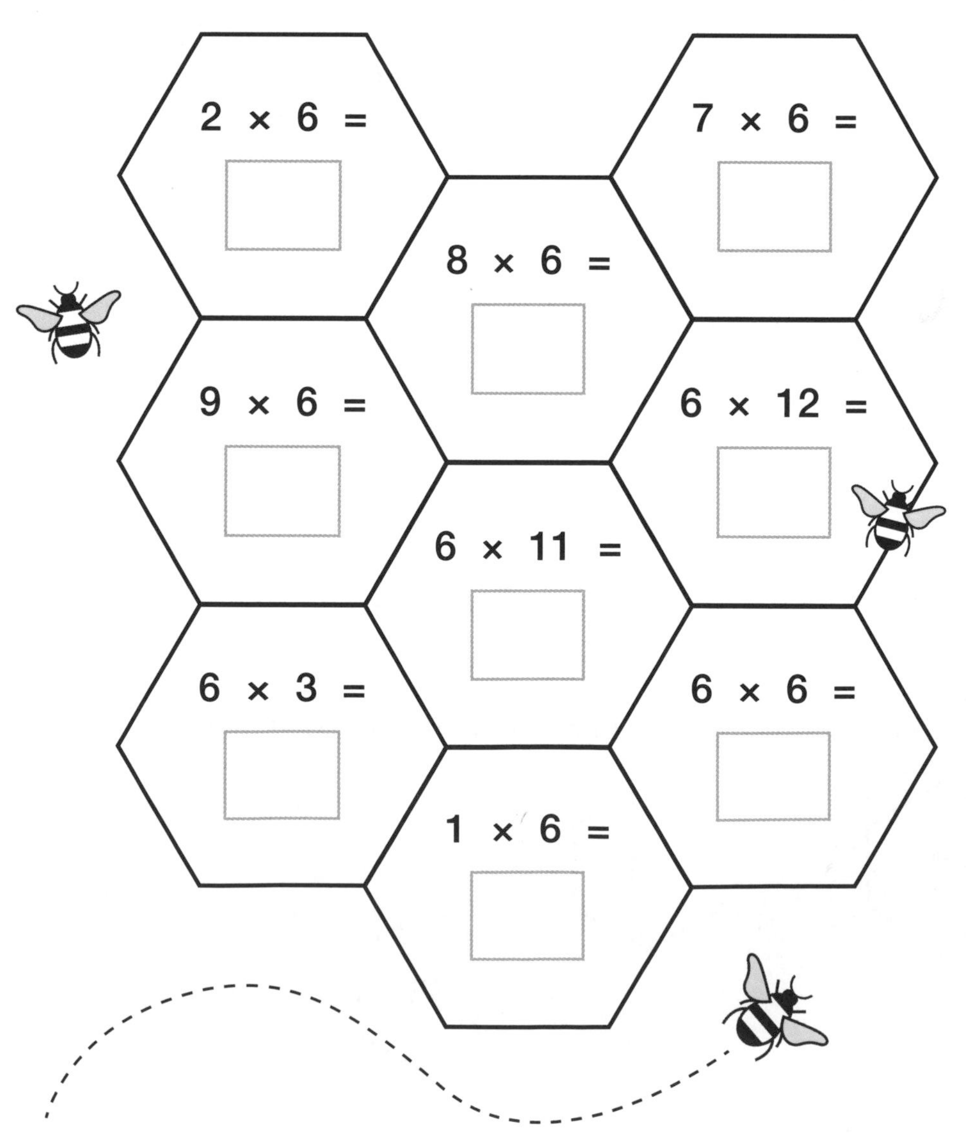

(D) Circle the multiples of 3 in the first row of numbers.
Circle the multiples of 6 in the second row of numbers. [10]

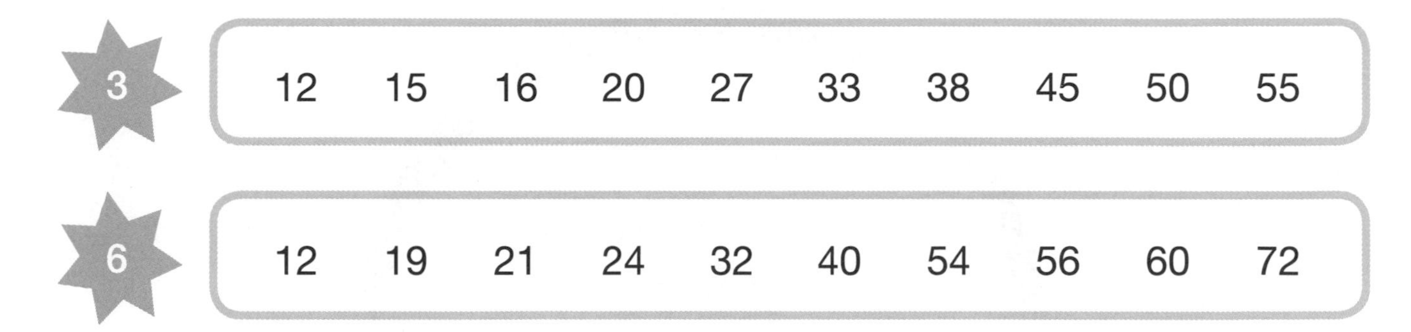

What is the lowest common multiple of 3 and 6 from your circled numbers? [1]

____

(E) Use the 6 times table to answer these word problems. [5]

1 A dog sanctuary gives each of its nine dogs six treats each.
How many treats did they give out altogether?

____

2 One bunch of flowers has six flowers in it. How many flowers are in five bunches of flowers?

____

3 A recipe makes six cupcakes. How many cupcakes will be made if the recipe is made three times?

____

4 An insect has 6 legs. There are 24 legs. How many insects are there?

____

5 Jenna got six pieces of homework every week for the whole term. The term was seven weeks long. How many pieces of homework did she get altogether?

____

Total marks

/ **16**

Ⓕ Target practice! How fast can you multiply the numbers in the ring by the number in the middle? Write your answers in the outer ring. One has been done for you. [9]

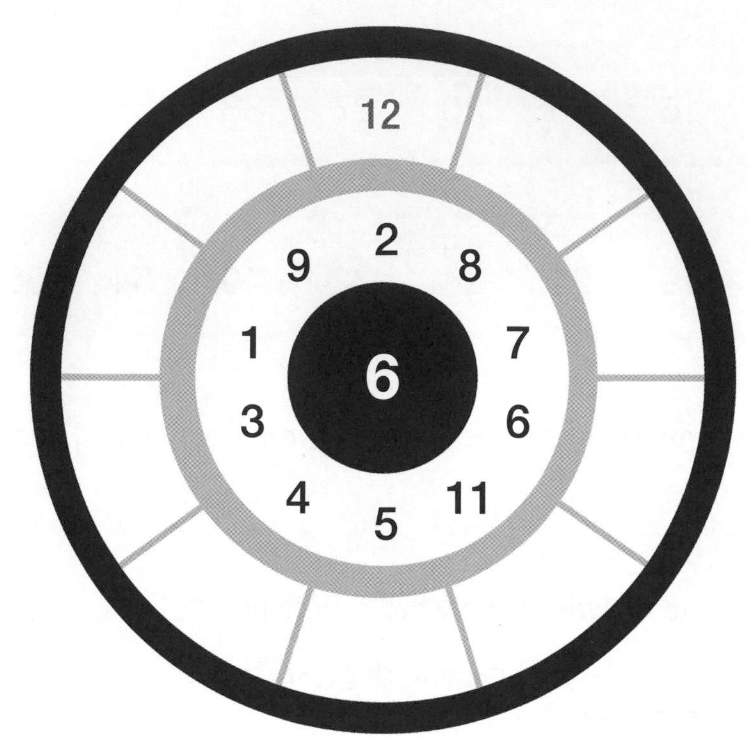

Ⓖ Fill in the missing numbers. [12]

5 × 6 = ☐	66 ÷ ☐ = 6
☐ ÷ 6 = 12	6 × 6 = ☐
6 × 0 = ☐	24 ÷ 6 = ☐
54 ÷ ☐ = 9	☐ × 6 = 18
☐ × 6 = 48	10 × 6 = ☐
72 ÷ 6 = ☐	☐ ÷ 6 = 1

Total marks ☐ / 21

The 11 times table

Pattern alert!

There is a very special pattern to the answers in the **11 times table** up to 9 × 11. Can you spot it?

Ⓐ Read and learn the 11 times table at the edge of the page. Now fold it under and fill in the missing numbers. Check your answers. [12]

1 × 11 = ☐	7 × 11 = ☐
2 × 11 = ☐	8 × 11 = ☐
3 × 11 = ☐	9 × 11 = ☐
4 × 11 = ☐	10 × 11 = ☐
5 × 11 = ☐	11 × 11 = ☐
6 × 11 = ☐	12 × 11 = ☐

Ⓑ Can you count in multiples of 11? Fill in the missing numbers. [7]

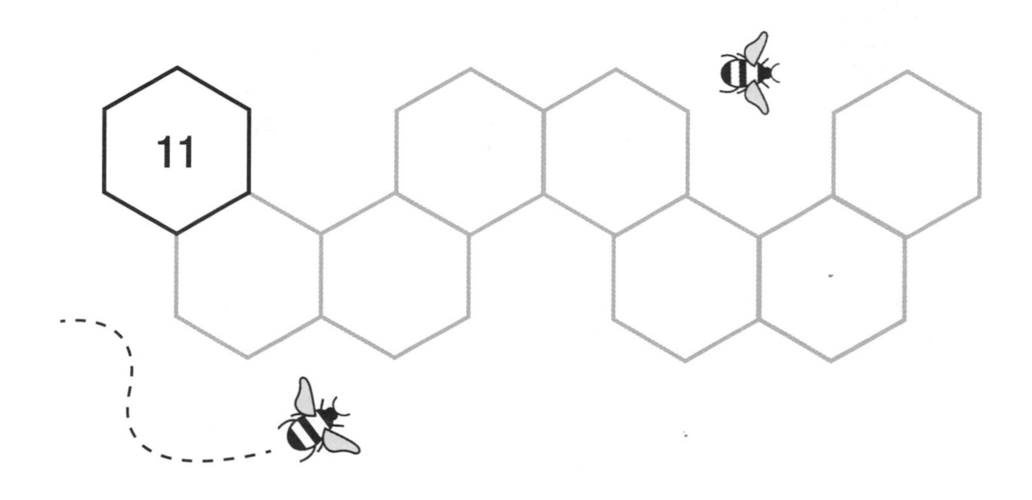

11
times table

1 × 11 = 11

2 × 11 = 22

3 × 11 = 33

4 × 11 = 44

5 × 11 = 55

6 × 11 = 66

7 × 11 = 77

8 × 11 = 88

9 × 11 = 99

10 × 11 = 110

11 × 11 = 121

12 × 11 = 132

Total marks

/19

11
times table

1 × 11 = 11

2 × 11 = 22

3 × 11 = 33

4 × 11 = 44

5 × 11 = 55

6 × 11 = 66

7 × 11 = 77

8 × 11 = 88

9 × 11 = 99

10 × 11 = 110

11 × 11 = 121

12 × 11 = 132

Total marks

/8

ⓒ Fold back the edge of this page and put the numbers in the number machines. Write the numbers that will come out. [8]

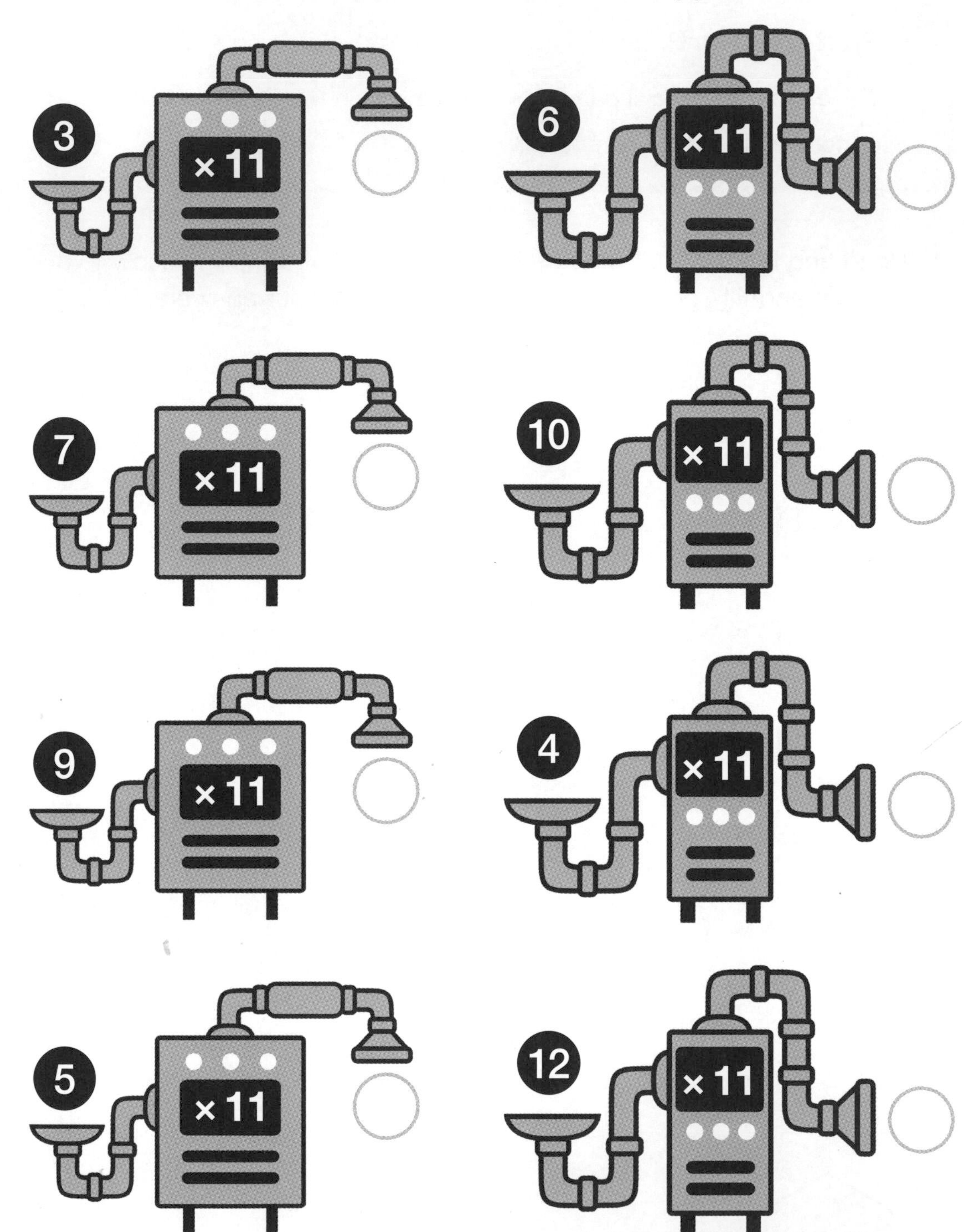

Helpful Hint

The **11 times table** helps you find groups of **11**.
Remember that there are different ways of saying groups of 11.

Example: **3** groups of **11** is the same as **3** lots of **11**,
3 multiplied by **11** and **3 × 11**.

(D) Fill in the missing numbers. [12]

88 = ☐ × 11	22 = ☐ × 2
33 = ☐ × 3	99 = ☐ × 11
55 = ☐ × 11	132 = ☐ × 11
77 = ☐ × 11	44 = ☐ × 4
121 = ☐ × 11	66 = ☐ × 6
11 = ☐ × 1	110 = ☐ × 11

(E) Answer these word problems. [4]

1 Six runners each ran eleven laps of the school track. How many laps of the school track were run altogether?

____

2 33 flapjacks were sold to 11 customers. If each customer bought the same number of flapjacks, how many did each customer buy?

____

3 A group of 11 litter-pickers each picked up 12 items of litter. How many items of litter were picked up in total?

____

4 44 buttons need to be sewn on to 11 coats. All the coats need the same number of buttons. How many buttons will be on each coat?

____

Total marks

/16

Helpful Hint

Remember, **multiplication** can help with **division**.

3 lots of 11 = 33

3 multiplied by 11 = 33

33 shared by 11 = 3

33 divided by 11 = 3

33 divided by 3 = 11

(F) Use the numbers in the box to fill in the blanks in the multiplications and divisions below. Some answers will be needed more than once. [16]

5	33	132	66	88	12	2
121	11	77	44	4	3	

$33 \div 11 =$ ☐	$7 \times 11 =$ ☐
$99 \div 9 =$ ☐	$44 \div 11 =$ ☐
$8 \times 11 =$ ☐	$11 \times 6 =$ ☐
$11 \times 11 =$ ☐	$22 \div 11 =$ ☐
$4 \times 11 =$ ☐	$110 \div 10 =$ ☐
$11 \times 3 =$ ☐	$132 \div 11 =$ ☐
$77 \div 7 =$ ☐	$55 \div 11 =$ ☐
$11 \times 12 =$ ☐	$1 \times 11 =$ ☐

Total marks ☐ / 16

Answers

Unit 1

(A) **2 times table**: 6, 2, 7, 2, 10, 10, 2, 6; **3 times table**: 21, 10, 4, 9, 36, 11, 9, 1

(B) **5 times table**: 35, 2, 5, 9, 60, 8, 15, 5; **10 times table**: 50, 30, 10, 6, 110, 10, 2, 120

(C) **2**: 6, 10, 20, 24, 30
5: 15, 20, 25, 45, 55
10: 20, 50, 70, 90, 120
20 is a multiple of 2, 5 and 10

(D) **1–8** 50, 80, 6, 9, 14, 5, 100, 5

(E) In order of the list given: 12, 18, 6, 10, 11, 2, 90, 4, 100, 20, 30, 25, 8, 5

Unit 2

(A) 4, 8, 12, 16, 20, 24, 28, 32, 36, 40, 44, 48

(B) 12, 16, 24, 32, 36

(C) 36, 20, 28, 32, 8, 48, 12

(D) 28 children, 20 tins, 44 slices

(E) 2, 3, 4, 5, 6, 9, 12

(F) 20, 48, 4, 4, 8, 2, 1, 9, 44, 6, 0, 5, 4, 40, 12, 1, 3, 24

Unit 3

(A) 8, 16, 24, 32, 40, 48, 56, 64, 72, 80, 88, 96

(B) 3, 8, 8, 12, 8, 6, 8, 7

(C) 24, 64, 8, 96, 40, 1, 7, 11, 32, 16, 6, 56, 12, 72

(D) 7 spiders, 8 apples, 40 words, 16 wings, 96 pens

(E) **4**: 12, 16, 20, 36, 48
8: 16, 32, 40, 56, 88
16 is a common multiple of 4 and 8

(F) From left to right: 24, 36, 16, 48, 72, 32, 6, 12, 3, 3, 6, 2

Unit 4

(A) 6, 12, 18, 24, 30, 36, 42, 48, 54, 60, 66, 72

(B) 12, 18, 30, 42, 48

(C) 12, 48, 42, 54, 66, 72, 18, 6, 36

(D) **3**: 12, 15, 27, 33, 45
6: 12, 24, 54, 60, 72
12 is the lowest common multiple

(E) **1–5** 54 treats, 30 flowers, 18 cupcakes, 4 insects, 42 pieces of homework

(F) Clockwise from top: 48, 42, 36, 66, 30, 24, 18, 6, 54

(G) 30, 72, 0, 6, 8, 12, 11, 36, 4, 3, 60, 6

Unit 5

(A) 11, 22, 33, 44, 55, 66, 77, 88, 99, 110, 121, 132

(B) 22, 33, 44, 55, 66, 77, 88

(C) 33, 66, 77, 110, 99, 44, 55, 132

(D) 8, 11, 5, 7, 11, 11, 11, 9, 12, 11, 11, 10

(E) 66 laps, 3 flapjacks, 132 items of litter, 4 buttons

(F) 3, 11, 88, 121, 44, 33, 11, 132, 77, 4, 66, 2, 11, 12, 5, 11

Unit 6

(A) 9, 18, 27, 36, 45, 54, 63, 72, 81, 90, 99, 108

(B)

Start 9	18	42	6	4	10
55	27	45	15	32	35
44	36	45	54	22	20
53	21	35	63	90	99
32	45	20	72	81	108
18	25	15	10	5	Exit

(C) 72, 63, 81, 45, 27, 108, 99

(D) 99 pockets, 81 songs, 54 pints of milk, 5 tables

(E) **x 9** clockwise from top: 54, 18, 0, 72, 90, 36
÷ 9 clockwise from top: 5, 9, 12, 1, 3

(F) 27, 81, 36, 3, 99, 1, 1, 2, 72, 8, 63, 5, 9, 54, 54, 9, 81, 12

Unit 7

(A) 6 x table: 30, 54, 12, 6, 4
8 x table: 32, 6, 7, 11, 96
9 x table: 72, 108, 6, 9, 9
11 x table: 77, 10, 11, 8, 66

(B) **6**: 6, 12, 18, 30, 36, 42, 48, 54, 60
4: 12, 28, 36, 44, 48
8: 48
9: 9, 18, 27, 36, 45, 54
11: 11, 22, 33, 44, 55

(C) 56, 99, 9, 66, 6

(D) 1, 9, 25, 64, 100, 4, 16, 36, 81, 121

(E) 1, 8, 27, 64, 125

(F)

×	4	6	8	9	11
1	4	6	8	9	11
4	16	24	32	36	44
5	20	30	40	45	55
8	32	48	64	72	88
3	12	18	24	27	33
6	24	36	48	54	66
9	36	54	72	81	99
2	8	12	16	18	22
12	48	72	96	108	132
7	28	42	56	63	77
11	44	66	88	99	121
10	40	60	80	90	110

Unit 8

(A) 12, 24, 36, 48, 60, 72, 84, 96, 108, 120, 132, 144

(B) 84, 144, 132, 96, 12, 36, 120

(C) £84, 48 legs, 3 PE lessons

(D) 3, 12, 5, 7, 4, 9, 11, 1

(E) 96, 9, 12, 132, 108, 12, 0, 10, 4, 12, 60, 3, 12, 5, 12, 120, 72, 12, 84, 9, 2, 3

Unit 9

(A) 7, 14, 21, 28, 35, 42, 49, 56, 63, 70, 77, 84

(B) 14, 35, 42, 56, 63, 84

(C) clockwise from top: 28, 21, 49, 35, 84, 63, 14, 56, 77, 42

(D) 7, 2, 7, 70, 9, 4, 5, 21, 63, 3, 28, 7, 42, 56, 77, 6, 49, 8, 84, 10

(E) 10, 1, 7, 7, 7, 6, 7, 12, 7, 2, 8, 11

(F) 35 days, £28, 8 strawberries, 9 children

Unit 10

(A) **From left to right**: 12, 21, 48, 45, 8, 72, 32, 35, 16, 60, 56, 12, 27, 30, 32, 99, 48, 3, 49, 24, 96, 55, 100, 54, 20, 6, 48, 64, 21, 16, 30, 18, 24, 72, 30, 27, 80, 84, 12, 9, 77, 36, 28, 18, 121

(B) **From left to right**: 2, 4, 7, 4, 7, 4, 8, 5, 10, 9, 8, 6, 8, 11, 9, 6, 10, 5, 7, 9, 7, 6, 3, 5, 4, 2, 5, 8, 12, 9, 5, 7, 10, 6, 7, 10, 3, 7, 3, 7, 4, 4, 9, 6, 8

(C) **From left to right**: 4, 24, 4, 24, 5, 63, 3, 49, 4, 8, 8, 36, 40, 10, 88, 9, 7, 9, 9, 25, 5, 64, 84, 7, 21, 9, 42, 9, 35, 5, 80, 11, 16, 18, 28, 7, 6, 42, 18, 16, 20, 12, 5, 9, 56

(D) **From left to right**: 4, 3, 54, 33, 64, 5, 7, 9, 8, 9, 35, 8, 5, 24, 12, 28, 9, 42, 56, 24, 2, 8, 25, 28, 12, 2, 6, 63, 12, 7, 8, 9, 40, 9, 8, 5, 9, 10, 125, 7, 66, 3, 10, 81, 54

The 9 times table

Pattern alert!

Look at the answers to the **9 times table**. Can you see how the tens **digits** go up one by one and the units go down? Try checking your answer by adding the digits in your answer together. They should add up to 9.

Example: 7 × 9 = 63, 6 + 3 = 9. Your answer is correct!

This works for all the **9 times table facts** apart from 11 x 9 = 99.

Ⓐ Read through the 9 times table at the edge of the page. Now cover it and fill in the missing numbers. Check your answers. [12]

1 × 9 = ☐	7 × 9 = ☐
2 × 9 = ☐	8 × 9 = ☐
3 × 9 = ☐	9 × 9 = ☐
4 × 9 = ☐	10 × 9 = ☐
5 × 9 = ☐	11 × 9 = ☐
6 × 9 = ☐	12 × 9 = ☐

Ⓑ Count on in 9s and colour the squares with the numbers in the 9 times table to trace a path through the number maze. [11]

Start 9	18	42	6	4	10
55	27	45	15	32	35
44	36	45	54	22	20
53	21	35	63	90	99
32	45	20	72	81	108
18	25	15	10	5	Exit

9
times table

1 × 9 = 9

2 × 9 = 18

3 × 9 = 27

4 × 9 = 36

5 × 9 = 45

6 × 9 = 54

7 × 9 = 63

8 × 9 = 72

9 × 9 = 81

10 × 9 = 90

11 × 9 = 99

12 × 9 = 108

Total marks

23

9
times table

1 × 9 = 9

2 × 9 = 18

3 × 9 = 27

4 × 9 = 36

5 × 9 = 45

6 × 9 = 54

7 × 9 = 63

8 × 9 = 72

9 × 9 = 81

10 × 9 = 90

11 × 9 = 99

12 × 9 = 108

Total marks

Helpful Hint

You can use your hands to help you with the **9 times table**.

Example: 3 × 9 = ?

Hold out your hands so that you have 10 **digits**.

Fold down the third finger of your left hand (because we are looking at 3 × 9).

Count the finger and thumb to the left of your folded over finger: 2

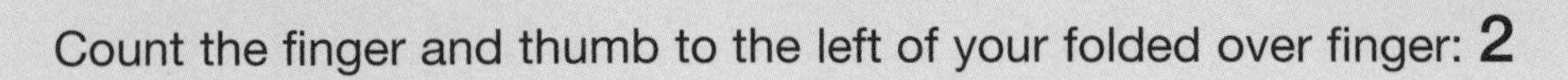

Count the fingers and thumb to the right of your folded over finger: 7

Put the **digits** together: 27   So 3 × 9 = 27

Ⓒ Fold back the edges of this page and balance the scales with the correct answers. One has been done for you. [7]

Ⓓ Use the 9 times table to answer these word problems. [4]

1 Several boxes were delivered to a supermarket. Each box contained nine packets of biscuits. How many packets of biscuits were there in eleven boxes?

____

2 Jess has nine albums on her music app. Each album has nine songs. How many songs are on her music app in total?

____

3 In one street, a milkman delivered six pints of milk each to nine homes. How many pints did he deliver in total?

____

4 At a dinner party there are nine guests per table. In total, there are forty-five guests. How many tables are there?

____

Ⓔ Target practice! How fast can you multiply or divide the numbers in the ring by the number in the middle? Fill in the missing numbers. [11]

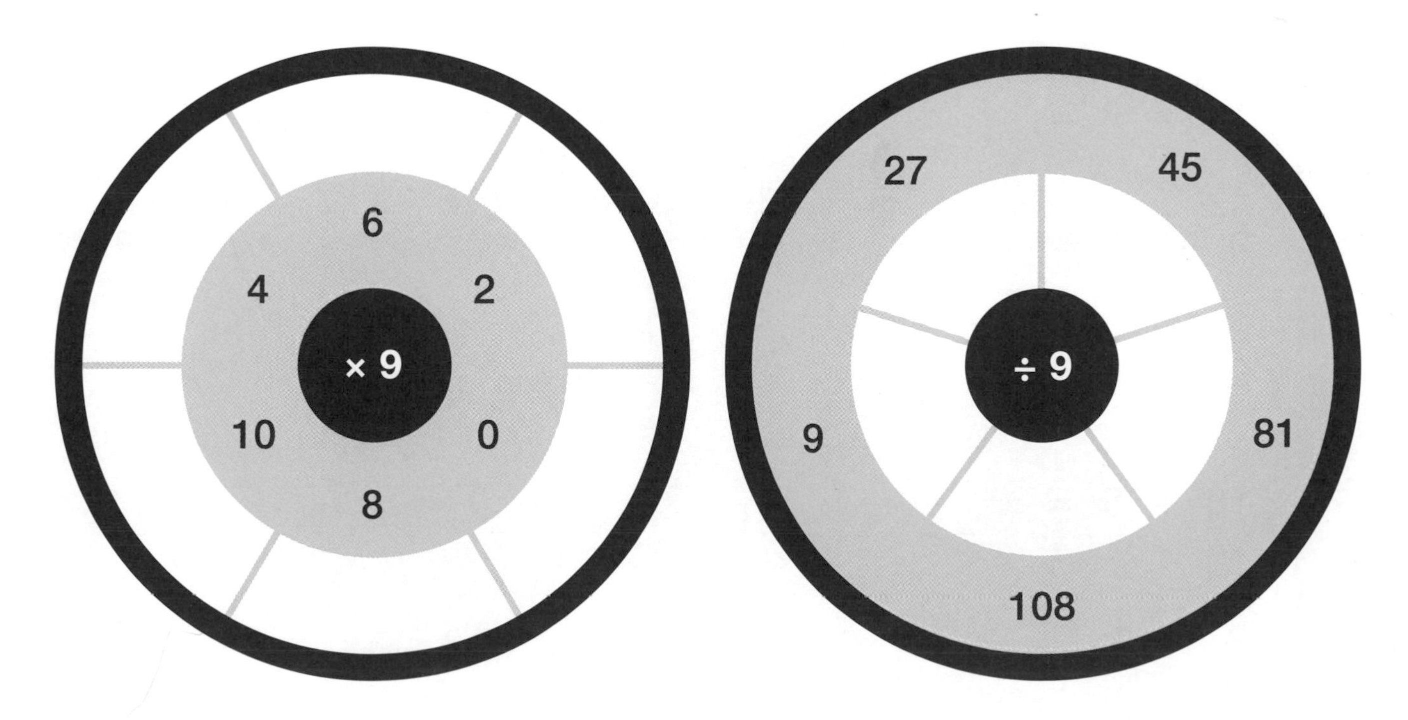

Total marks

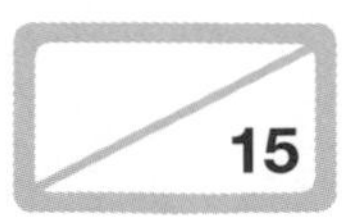

Helpful Hint

Remember that **multiplication** and **division** are connected.

$4 \times 9 = 36$ | $36 \div 9 = 4$

4 groups of 9 is 36 | 36 shared between 9 is 4

4 multiplied by 9 = 36 | 36 divided by 9 = 4

(F) Fill in the missing numbers. [18]

$9 \times 3 =$ ☐	$72 \div 9 =$ ☐
☐ $\div 9 = 9$	$7 \times 9 =$ ☐
$9 \times 4 =$ ☐	$45 \div 9 =$ ☐
$27 \div$ ☐ $= 9$	☐ $\times 2 = 18$
$11 \times 9 =$ ☐	$6 \times 9 =$ ☐
$9 \div 9 =$ ☐	☐ $\div 6 = 9$
$9 \times$ ☐ $= 9$	$63 \div$ ☐ $= 7$
$18 \div 9 =$ ☐	$9 \times 9 =$ ☐
$8 \times 9 =$ ☐	$108 \div$ ☐ $= 9$

Total marks

/18

Test your skills: mixed times tables

Ⓐ How well do you remember these tables? Fill in the missing numbers. [20]

6 × table

5 × 6 = ☐

9 × 6 = ☐

☐ × 6 = 72

3 × ☐ = 18

☐ × 6 = 24

8 × table

4 × 8 = ☐

☐ × 8 = 48

8 × ☐ = 56

☐ × 8 = 88

12 × 8 = ☐

9 × table

8 × 9 = ☐

12 × 9 = ☐

☐ × 9 = 54

7 × ☐ = 63

☐ × 9 = 81

11 × table

7 × 11 = ☐

☐ × 11 = 110

5 × ☐ = 55

☐ × 11 = 88

6 × 11 = ☐

Total marks ☐ / 20

Ⓑ Sort the numbers in the cloud into the table. Some multiples will go into more than one column. Give yourself one mark for each correct column. [5]

36 28 22 44 42 18 12 60 54 ~~24~~ 11 9 6 45 33 55 48 30 27

Multiples of 6	Multiples of 4	Multiples of 8	Multiples of 9	Multiples of 11
24	24	24		

Ⓒ Write the answers to these questions. [5]

1 Multiply seven by eight. ____________

2 Find the product of 11 and nine. ____________

3 What is 54 shared between six? ____________

4 Multiply six and 11. ____________

5 Divide 48 equally between eight. ____________

Total marks

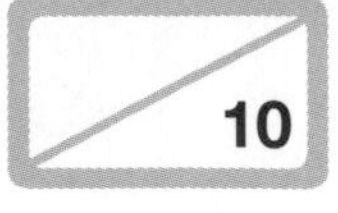

Helpful Hint

When a number is multiplied by itself the **product** is called a **square number**.

Example: $3 \times 3 = 9$. This can also be written as $3^2 = 9$. 9 is a **square number**.

The **square number** of an even number is always even.
The **square number** of an odd number is always odd.

When a number is multiplied by itself and the answer is multiplied by the same number again, it is called a **cube number**.

Example: $2 \times 2 \times 2 = 8$ or $2^3 = 8$. 8 is a **cube number**.

D Answer these as quickly as possible. The answers are all square numbers. [10]

$1 \times 1 =$ ☐	$2^2 =$ ☐
$3 \times 3 =$ ☐	$4 \times 4 =$ ☐
$5^2 =$ ☐	$6 \times 6 =$ ☐
$8 \times 8 =$ ☐	$9^2 =$ ☐
$10 \times 10 =$ ☐	$11 \times 11 =$ ☐

E Fill in the missing numbers. The answers are all cube numbers. [5]

$1^3 = 1 \times 1 \times 1 =$ ☐ $2^3 = 2 \times 2 \times 2 =$ ☐

$3^3 =$ ☐ $\times$ ☐ $\times$ ☐ $=$ ☐

$4^3 =$ ☐ $\times$ ☐ $\times$ ☐ $=$ ☐

$5^3 =$ ☐ $\times$ ☐ $\times$ ☐ $=$ ☐

Total marks /15

Ⓕ How quickly can you fill in this multiplication grid? Give yourself one mark for completing each row. The first one has been done for you. [11]

×	4	6	8	9	11
1	4	6	8	9	11
4					
5					
8					
3					
6					
9					
2					
12					
7					
11					
10					

Total marks

/11

The 12 times table

Pattern alert!

Like all even **times tables**, all the numbers in the **12 times table** are even numbers. They end with 0, 2, 4, 6 or 8.

If you **multiply** a number by 12 it is the same as multiplying that number by 10 and multiplying it by 2 and adding those **products** together.

Example: To multiply 6 by 12 →

$$\begin{aligned} 6 \times 10 &= 60 + \\ 6 \times 2 &= \underline{12} \\ & \quad 72 \end{aligned}$$

The answers to the **12 times table** are **double** that of the **6 times table**.

(A) Read and learn the 12 times table at the edge of the page.
Now cover it and fill in the missing numbers. Check your answers. [12]

1 × 12 =		7 × 12 =	
2 × 12 =		8 × 12 =	
3 × 12 =		9 × 12 =	
4 × 12 =		10 × 12 =	
5 × 12 =		11 × 12 =	
6 × 12 =		12 × 12 =	

12
times table

1 × 12 = 12

2 × 12 = 24

3 × 12 = 36

4 × 12 = 48

5 × 12 = 60

6 × 12 = 72

7 × 12 = 84

8 × 12 = 96

9 × 12 = 108

10 × 12 = 120

11 × 12 = 132

12 × 12 = 144

Total marks

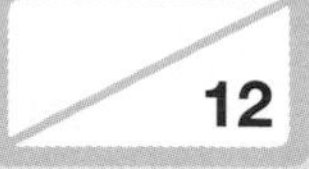

Unit 8

12
times table

1 × 12 = 12

2 × 12 = 24

3 × 12 = 36

4 × 12 = 48

5 × 12 = 60

6 × 12 = 72

7 × 12 = 84

8 × 12 = 96

9 × 12 = 108

10 × 12 = 120

11 × 12 = 132

12 × 12 = 144

Total marks

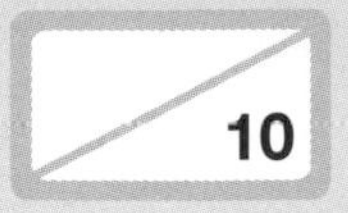

Ⓑ Fold back the edge of this page and balance the scales with the correct answers. One has been done for you. [7]

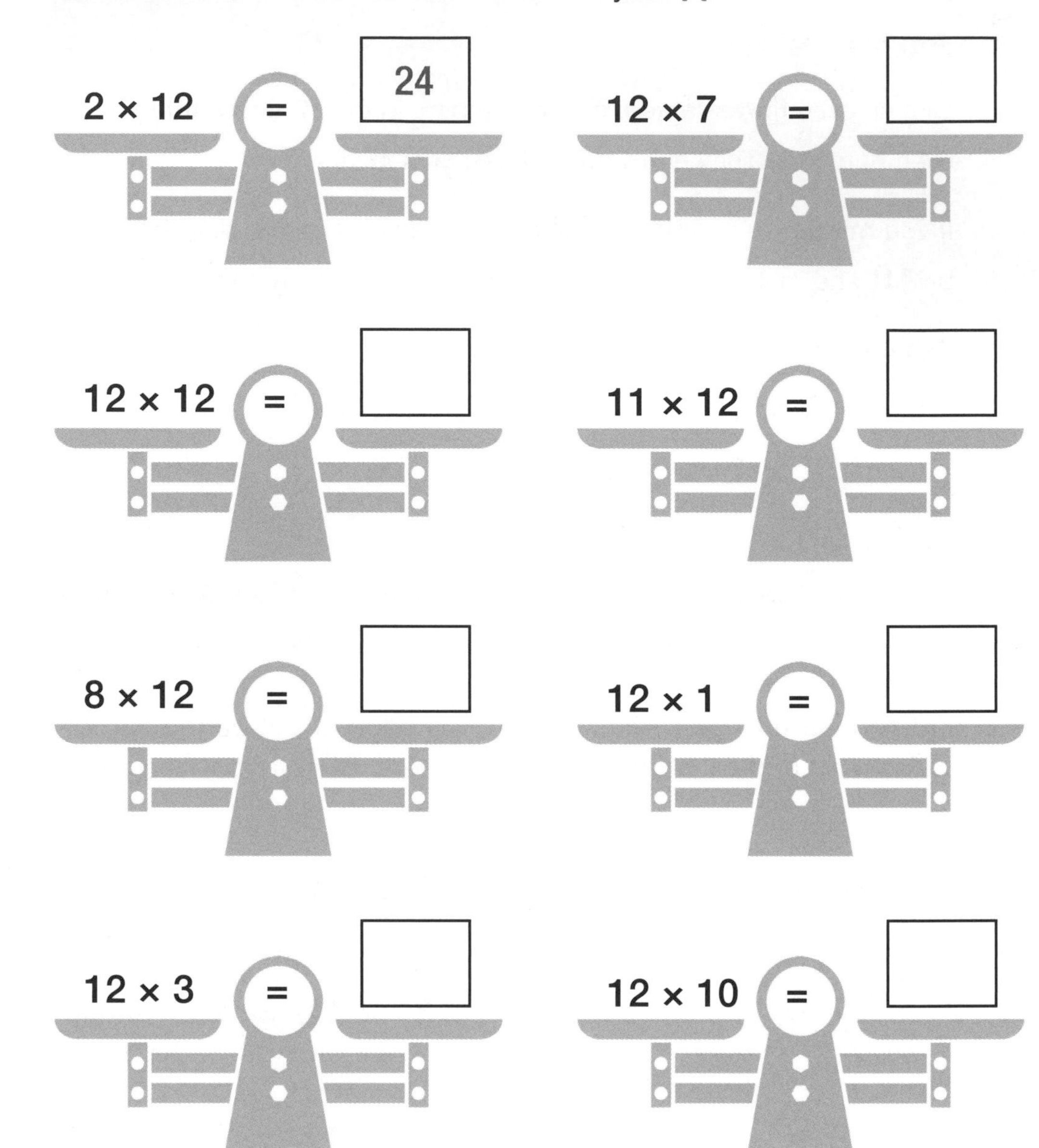

Ⓒ Use the 12 times table to answer these word problems. [3]

1 Grandad gave his seven grandchildren £12 each. How much money did he hand out altogether?

____

2 How many legs do twelve cats have?

____

3 In twelve weeks Tuhil has thirty-six PE lessons. How many PE lessons does he have in one week?

____

Helpful Hint

You can work out a **division** problem by thinking of the matching **multiplication fact**.

Example: $2 \times 12 = 24$ $24 \div 12 = 2$ $24 \div 2 = 12$

(D) Put the numbers through the number machine. Write the numbers that come out. [8]

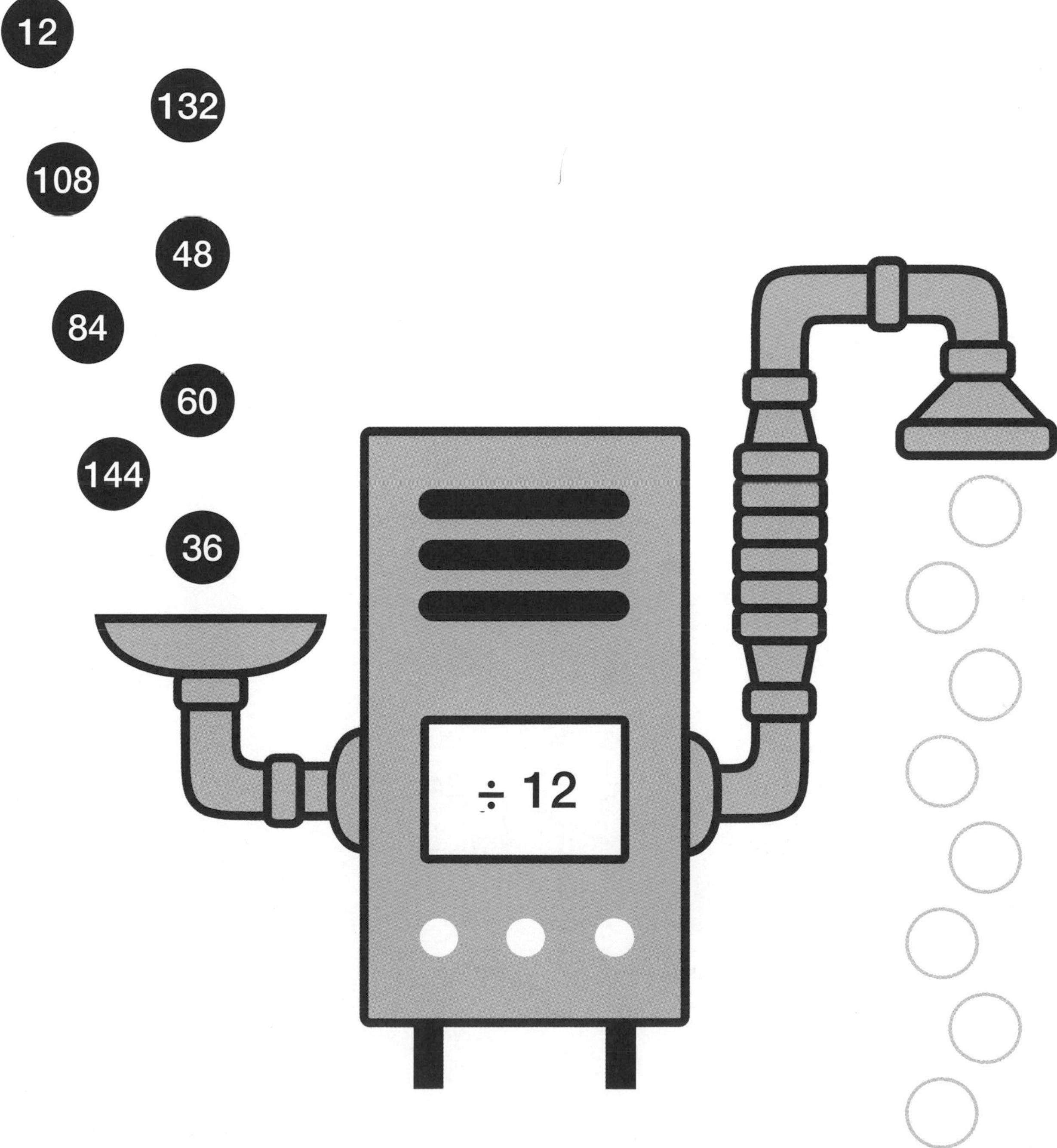

Total marks

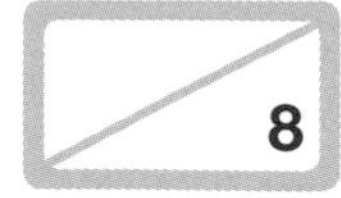

Ⓔ Fill in the missing numbers. [22]

$12 \times 8 = \square$	$36 \div \square = 12$
$108 \div 12 = \square$	$7 \times \square = 84$
$\square \times 12 = 144$	$60 \div 12 = \square$
$\square \div 12 = 11$	$\square \times 2 = 24$
$9 \times 12 = \square$	$10 \times 12 = \square$
$48 \div 4 = \square$	$\square \div 6 = 12$
$12 \times 0 = \square$	$12 \div \square = 1$
$120 \div \square = 12$	$\square \div 12 = 7$
$12 \times \square = 48$	$12 \times \square = 108$
$144 \div 12 = \square$	$24 \div 12 = \square$
$5 \times 12 = \square$	$\square \times 12 = 36$

Total marks

/22

The 7 times table

Pattern alert!

There may not be much of a pattern to the **7 times table** but you know 11 of the facts from all of the tables you have already learnt!

The only **fact** you haven't learnt yet is $7 \times 7 = 49$

Just as the **12 times table** can be split up into **times** 10 then **times** 2, the **7 times table** can be split up into **times** 5 and **times** 2 and the answers added together.

Example: $8 \times 7 = ?$

$8 \times 5 = 40$

$8 \times 2 = 16$

$40 + 16 = 56$

(A) Read and learn the 7 times table at the edge of the page. Now cover it and fill in the missing numbers. Check your answers. [12]

$1 \times 7 =$ ☐	$7 \times 7 =$ ☐
$2 \times 7 =$ ☐	$8 \times 7 =$ ☐
$3 \times 7 =$ ☐	$9 \times 7 =$ ☐
$4 \times 7 =$ ☐	$10 \times 7 =$ ☐
$5 \times 7 =$ ☐	$11 \times 7 =$ ☐
$6 \times 7 =$ ☐	$12 \times 7 =$ ☐

7
times table

1 × 7 = 7

2 × 7 = 14

3 × 7 = 21

4 × 7 = 28

5 × 7 = 35

6 × 7 = 42

7 × 7 = 49

8 × 7 = 56

9 × 7 = 63

10 × 7 = 70

11 × 7 = 77

12 × 7 = 84

Total marks

/12

7
times table

1 × 7 = 7

2 × 7 = 14

3 × 7 = 21

4 × 7 = 28

5 × 7 = 35

6 × 7 = 42

7 × 7 = 49

8 × 7 = 56

9 × 7 = 63

10 × 7 = 70

11 × 7 = 77

12 × 7 = 84

Total marks

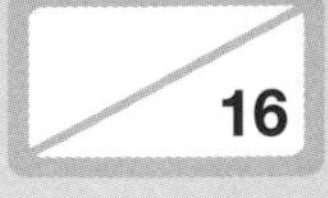

(B) Fold back the edge of this page and count forwards and backwards in 7s. Fill in the missing numbers in this sequence. [6]

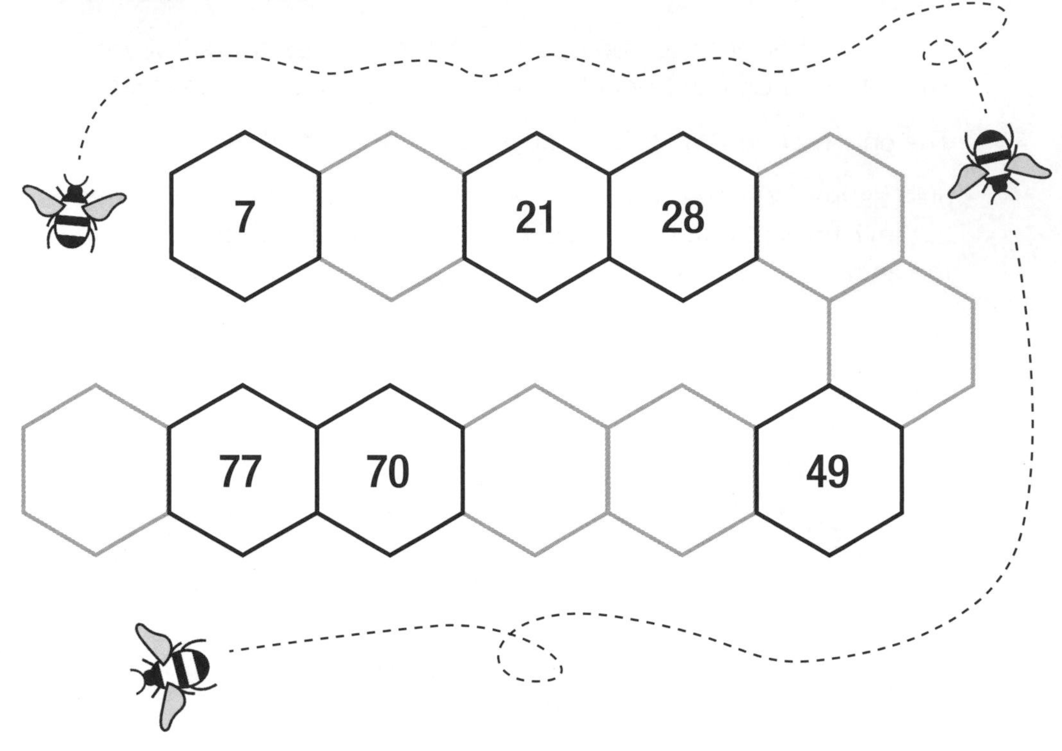

(C) Target practice. How fast can you multiply the numbers in the ring by the number in the middle? [10]

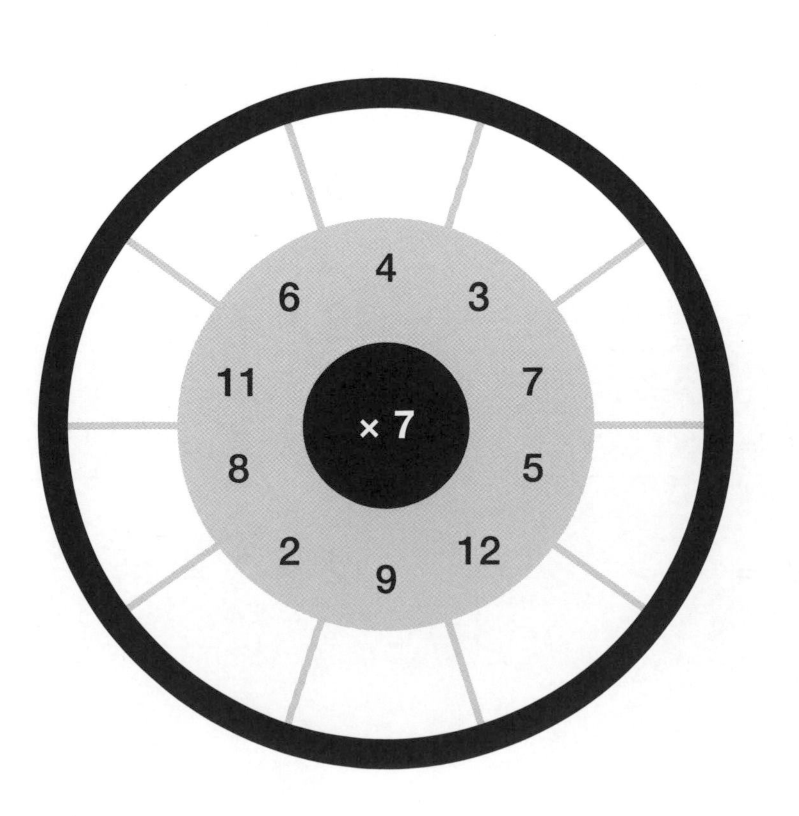

Ⓓ Use the numbers in the box to fill in the answers to the multiplications or divisions. Watch out, answers may be needed more than once. [20]

5	77	28	56	7	8	4	63	84
42	9	6	21	70	2	49	3	10

$49 \div 7 =$ ☐ $4 \times 7 =$ ☐

$14 \div 7 =$ ☐ $7 \div 1 =$ ☐

$84 \div 12 =$ ☐ $7 \times 6 =$ ☐

$10 \times 7 =$ ☐ $8 \times 7 =$ ☐

$63 \div 7 =$ ☐ $11 \times 7 =$ ☐

$28 \div 7 =$ ☐ $42 \div 7 =$ ☐

$35 \div 7 =$ ☐ $7 \times 7 =$ ☐

$7 \times 3 =$ ☐ $56 \div 7 =$ ☐

$9 \times 7 =$ ☐ $12 \times 7 =$ ☐

$21 \div 7 =$ ☐ $70 \div 7 =$ ☐

Total marks

/20

(E) Fill in the missing numbers. [12]

70 =		× 7		28 =		× 4
7 =		× 7		84 =		× 7
63 =		× 9		21 =		× 3
49 =		× 7		14 =		× 7
35 =		× 5		56 =		× 7
42 =		× 7		77 =		× 7

(F) Use the 7 times table to answer these word problems. [4]

1 How many days are there in five weeks?

2 A toy car costs £7. How much do four of these toy cars cost?

3 Sam picked 56 strawberries. When he got home he shared them equally between seven people. How many strawberries did they each get?

4 63 children were going on a school camping trip. They were split equally into seven tents. How many children were in each tent?

Total marks

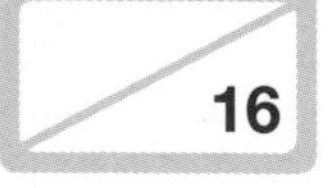

Time yourself!

Ⓐ How quickly can you complete these questions? Use a pencil so you can repeat the challenge and try and beat your time. [45]

2 × 6 = ______	7 × 3 = ______	6 × 8 = ______
9 × 5 = ______	8 × 1 = ______	12 × 6 = ______
8 × 4 = ______	5 × 7 = ______	4 × 4 = ______
6 × 10 = ______	8 × 7 = ______	6 × 2 = ______
3 × 9 = ______	5 × 6 = ______	4 × 8 = ______
9 × 11 = ______	8 × 6 = ______	3 × 1 = ______
7 × 7 = ______	6 × 4 = ______	12 × 8 = ______
11 × 5 = ______	10 × 10 = ______	6 × 9 = ______
4 × 5 = ______	2 × 3 = ______	4 × 12 = ______
8 × 8 = ______	3 × 7 = ______	8 × 2 = ______
3 × 10 = ______	6 × 3 = ______	2 × 12 = ______
8 × 9 = ______	6 × 5 = ______	9 × 3 = ______
10 × 8 = ______	7 × 12 = ______	4 × 3 = ______
1 × 9 = ______	7 × 11 = ______	6 × 6 = ______
7 × 4 = ______	9 × 2 = ______	11 × 11 = ______

Time: [ ]

Total marks [/45]

(B) How quickly can you answer these questions? Ask someone to time you! Use a pencil so you can try again and see if you can beat your time. [45]

10 ÷ 5 = ______

24 ÷ 6 = ______

40 ÷ 5 = ______

63 ÷ 7 = ______

24 ÷ 3 = ______

54 ÷ 9 = ______

7 ÷ 1 = ______

36 ÷ 6 = ______

28 ÷ 7 = ______

16 ÷ 2 = ______

10 ÷ 2 = ______

42 ÷ 7 = ______

9 ÷ 3 = ______

77 ÷ 11 = ______

108 ÷ 12 = ______

32 ÷ 8 = ______

56 ÷ 8 = ______

35 ÷ 7 = ______

48 ÷ 6 = ______

66 ÷ 6 = ______

80 ÷ 8 = ______

18 ÷ 2 = ______

12 ÷ 4 = ______

6 ÷ 3 = ______

144 ÷ 12 = ______

35 ÷ 5 = ______

84 ÷ 12 = ______

63 ÷ 9 = ______

4 ÷ 1 = ______

18 ÷ 3 = ______

21 ÷ 3 = ______

48 ÷ 12 = ______

50 ÷ 5 = ______

12 ÷ 2 = ______

36 ÷ 4 = ______

15 ÷ 3 = ______

42 ÷ 6 = ______

40 ÷ 8 = ______

60 ÷ 12 = ______

72 ÷ 8 = ______

20 ÷ 2 = ______

90 ÷ 9 = ______

27 ÷ 9 = ______

16 ÷ 4 = ______

64 ÷ 8 = ______

Time: ______

Total marks ___ / 45

Unit 10

Ⓒ How fast are you? Use a pencil so this page can be used again! [45]

$20 \div 5 =$ ______

$3 \times 8 =$ ______

$36 \div 12 =$ ______

$48 \div 6 =$ ______

$5 \times 8 =$ ______

$3^2 =$ ______

$54 \div 6 =$ ______

$8 \times 8 =$ ______

$3 \times 7 =$ ______

$81 \div 9 =$ ______

$8 \times 10 =$ ______

$6 \times 3 =$ ______

$48 \div 8 =$ ______

$4^2 =$ ______

$25 \div 5 =$ ______

$6 \times 4 =$ ______

$45 \div 9 =$ ______

$7 \times 7 =$ ______

$24 \div 3 =$ ______

$50 \div 5 =$ ______

$28 \div 4 =$ ______

$5^2 =$ ______

$7 \times 12 =$ ______

$9 \div 1 =$ ______

$5 \times 7 =$ ______

$99 \div 9 =$ ______

$4 \times 7 =$ ______

$6 \times 7 =$ ______

$10 \times 2 =$ ______

$27 \div 3 =$ ______

$44 \div 11 =$ ______

$9 \times 7 =$ ______

$2^2 =$ ______

$9 \times 4 =$ ______

$8 \times 11 =$ ______

$72 \div 8 =$ ______

$30 \div 6 =$ ______

$56 \div 8 =$ ______

$7 \times 6 =$ ______

$15 \div 3 =$ ______

$8 \times 2 =$ ______

$35 \div 5 =$ ______

$9 \times 2 =$ ______

$72 \div 6 =$ ______

$8 \times 7 =$ ______

Time:

Total marks /45

(D) Are you getting quicker? [45]

2^2 = ______	24 ÷ 8 = ______	9 × 6 = ______
3 × 11 = ______	4^3 = ______	45 ÷ 9 = ______
70 ÷ 10 = ______	18 ÷ 2 = ______	3^3 = ______
72 ÷ 8 = ______	7 × 5 = ______	24 ÷ 3 = ______
55 ÷ 11 = ______	4 × 6 = ______	48 ÷ 4 = ______
7 × 4 = ______	45 ÷ 5 = ______	7 × 6 = ______
8 × 7 = ______	12 × 2 = ______	18 ÷ 9 = ______
8 ÷ 1 = ______	5^2 = ______	4 × 7 = ______
72 ÷ 6 = ______	14 ÷ 7 = ______	48 ÷ 8 = ______
9 × 7 = ______	132 ÷ 11 = ______	49 ÷ 7 = ______
16 ÷ 2 = ______	36 ÷ 4 = ______	4 × 10 = ______
3^2 = ______	2^3 = ______	15 ÷ 3 = ______
27 ÷ 3 = ______	20 ÷ 2 = ______	5^3 = ______
56 ÷ 8 = ______	6 × 11 = ______	36 ÷ 12 = ______
80 ÷ 8 = ______	9 × 9 = ______	6 × 9 = ______

Time: [ ]

Total marks ___ / 45

Key words

cube number the product of three identical numbers, for example *4 x 4 x 4 = 64 or* 4^3 *= 64*

digit any of the numerals on their own or when used to make another number, for example *digits: 0, 1, 2, 3, 4, 5, 6, 7, 8, 9* or used to make another number: *279*

divide to share an amount equally between groups, for example *30 chocolates shared between 5 children is 6 chocolates for each child or 30 ÷ 5 = 6*

÷ the symbol we use to mean 'divided by'

division sharing things equally or grouping things into sets of the same size, for example *8 ÷ 2 = 4 = what is 8 shared equally between 2? = how many groups of 2 are in 8?*

double/doubling to multiply a number by 2 or to add two numbers together that are the same, for example *12 x 2 = 24* and *12 + 12 = 24*

half/halving to divide a number by 2 or to split into two equal halves, for example *24 ÷ 2 = 12*

multiplication adding lots of the same number together, for example *3 x 4 = 3 + 3 + 3 + 3 = 12*

× the symbol we use to mean 'multiplied by'

multiplication fact how we write out the times tables, for example *1 x 5, 2 x 5, 3 x 5* and so on

multiple the number made by multiplying two numbers (the multiplication fact), for example *2 x 5 = 10, 3 x 5 = 15, 4 x 5 = 20* so *10, 15* and *20* are all multiples of 5

multiply to times two numbers together

product the answer you get by multiplying numbers together, for example *5 x 6 = 30*

square number the product of two identical numbers, for example *4 x 4 = 16* or 4^2 *= 16*

times to multiply two numbers together

times table a list of multiplication facts and multiples for a particular number, usually up to 12, for example the 2 times table: *1 x 2 = 2, 2 x 2 = 4, 3 x 2 = 6* and so on

Progress chart

How did you do? Fill in your score. Shade the matching boxes so that you can see how well you are doing in the different units.

50% 100%

Unit 1, p3 Score: ______ /16

Unit 1, p4 Score: ______ /16

Unit 1, p5 Score: ______ /24

Unit 1, p6 Score: ______ /14

Unit 2, p7 Score: ______ /17

Unit 2, p8 Score: ______ /7

Unit 2, p9 Score: ______ /10

Unit 2, p10 Score: ______ /18

Unit 3, p11 Score: ______ /12

Unit 3, p12 Score: ______ /22

Unit 3, p13 Score: ______ /16

Unit 3, p14 Score: ______ /12

Unit 4, p15 Score: ______ /17

Unit 4, p16 Score: ______ /9

Unit 4, p17 Score: ______ /16

Unit 4, p18 Score: ______ /21

Unit 5, p19 Score: ______ /19

Unit 5, p20 Score: ______ /8

Unit 5, p21 Score: ______ /16

Unit 5, p22 Score: ______ /16

50% 100%

Unit 6, p27 Score: ______ /23

Unit 6, p28 Score: ______ /7

Unit 6, p29 Score: ______ /15

Unit 6, p30 Score: ______ /18

Unit 7, p31 Score ______ /20

Unit 7, p32 Score: ______ /10

Unit 7, p33 Score: ______ /15

Unit 7, p34 Score: ______ /11

Unit 8, p35 Score: ______ /12

Unit 8, p36 Score: ______ /10

Unit 8, p37 Score: ______ /8

Unit 8, p38 Score: ______ /22

Unit 9, p39 Score: ______ /12

Unit 9, p40 Score: ______ /16

Unit 9, p41 Score: ______ /20

Unit 9, p42 Score: ______ /16

Unit 10:

A	1st time:	2nd time:	3rd time:
B	1st time:	2nd time:	3rd time:
C	1st time:	2nd time:	3rd time:
D	1st time:	2nd time:	3rd time: